PARANORMAL QUESTIONS OF THE DAY BOOK 2

2020 VERSION WITH ALL NEW QUESTIONS CRYPTID EDITION

By Author Eric Perry

INTRODUCTION

This book is about cryptid creatures we hear about from stories handed down by generations. And eyewitness encounters.

This book will give you an insight to what people believe in. These creatures many believe to be real or hoaxes.

I have compiled a lot of answers from REDDIT, FACEBOOK, LINKEDIN, and INSTAGRAM.

Do you believe in the Lochness Monster, or the Yeti in the Himalayan Mountains, or the legend of Sasquatch in the northwest wilderness of the United States?

The study of Cryptozoology is an amazing science. New and undiscovered creatures are being found all over the world.

So sit back and enjoy this amazing and look into what really goes bump in the deep, dark forests, oceans or even the jungles of our world.

Are you a believer or just curious what's out there?

I AM A BELIEVER

ABOUT THE AUTHOR

Hello, My name is Eric Perry and I am the founder of a team called, Haunted in New England. I have always believed in the paranormal but now I have opened my eyes to more than just ghosts.

In 2017 when I was in Chesterfield, NH I took some of my friends to a haunted location known as Madame Sherri's Castle. It was dusk and we were wrapping up for the day. All of a sudden, we heard branches breaking and heard loud footsteps. We found a large footprint earlier in the day. But I didn't think anything about it. As we shined a light towards a bush, a large 8ft, hairy creature stood up! It smelled bad and had red eyes and let out an ungodly yell. We encountered BIGFOOT! It made me a believer .

This is my fifth book as an author. I am a proud father of 3 amazing kids, Logan, Aliyah, and my oldest, Eric Jr. I am a 3-time paranormal award winner and an international award winner in education. I can also be seen on Amazon Video with my team, Haunted in New England and more recently on the Travel Channel Shows, Haunted Hospitals and Paranormal 911. My hope is that this book opens possibilities to open your mind and see how undiscovered this world is until we explore our world.

THE LEGEND BIGFOOT

The legend is known by many names, Sasquatch, Bigfoot, Woldman and many others. The stories of this creature dates back to early Native American times. Bigfoot is known to stand 6ft to 7ft tall with many different colored hairs. Some people report a very pungent smell with glowing red eyes. The scientific community is sorting out some facts from the hoaxes.

The piece of evidence in the study of this creature is the Patterson-Gimlin film of 1967 in Bluff creek. The film was only a min long, but it showed a female Bigfoot walking away from the men as they filmed her. Some people today try and disprove this footage. Do you believe in Bigfoot ? I do.

You see back in 2013, in Parsonsfield, Me, I had an encounter with 2 Bigfoot creatures. It was a full moon night and it was a Friday the 13th at night as well. We were investigating the seminary for paranormal activity which we had a lot of evidence for. It was about 12 am we were heading back to command, when we heard a loud yelling sound and then heard branches break. As I turned the corner I saw an 8ft hairy beast with his hand on the shed.

I froze in my spot and made eye contact with it so did the rest of the crew.We heard a branch break behind then a yell again it was gone . Do i believe i saw bigfoot i do 100%

Now i got some amazing responses from the public on this and here are some of the best comments i received. No names are used.

Kind of. Don't understand why no one has ever found a bigfoot skeleton or individual bones somewhere.

Yes I believe in the yeti just as I do mermaids cause we have only explored so much of this planet's species so it's entirely possible.

Real. Both yeti and bigfoot exist but are not primates or any elusive species native to Earth. There is more to this than meets the eye, see Steve from how to hunt on YouTube for more information. Dig deep and you will find the uncomfortable truths.

Well to make the Bigfoot an endangered species I have to bring in a specimen dead or alive so I don't want to kill the Bigfoot.

It wouldn't surprise me if they are. So many areas of our wilderness and oceans haven't been explored fully yet.

Whether you believe in Bigfoot or not it's up to you to decide if the creature is real or fake.

The North American Wendigo

Do you believe in the Native American Wendigo

Real or fake?

#HAUNTEDINNEWENGLAND

The North American Wendigo is in many Native American folklore and handed down to generations. The Native American legend of the Wendigo is clouded in some truths and just stories.

The Wendigo is a mythological man-eating creature or an evil spirit. The folklore of the Wendigo comes from the First Nations Algonquin Tribes. The Wendigo is also known as a Flesheater and appears as a monster with the characteristics of a human.

When I started doing the research on Wendigo I was not sure what I was going to find. All the articles I read about the Wendigo seemed to be tales of flesh eating creatures. So was this real or a folklore? I asked here for some of the responses.

From what I've read, Natives never describe Wendigos with antlers. They were described as extremely tall skinny creatures, with tight skin that was shredded or falling apart, matted wild hair, and an exposed ribcage.

I have been researching on and off for years for a cryptid my husband, my son, and myself saw late at night on the plains of Northern Montana in the middle of icy snowy winter on Hwy 87 north of Fort Benton, MT.

It was an 8 foot tall creature, incredibly skinny long legs, very long wild unkempt hair (almost like a sea of tiny dreadlocks), dressed in flowing strips of sheet that moved with the wind. Hundreds of translucent white strips maybe 7" in length x 2" width undulating from its body. It was colorless, sorta glowy, and took on the color of the snowy bleak B&W environment. It stood at the side of the road completely still until our car came right to it in passing and then it took off in a jogging run. We all knew it was paranormal immediately, were tongue tied, and confused. My husband would not turn the car around and sped up to get away instead.

This creature has haunted me for years. I can find no info or images of anything that looks like it. Although, two other Redditors have described it exactly as I have on other forums. My son (registered Chippewa Cree) described it to some of his friends that live on the Browning Montana reservation, and they said it was a Skinwalker. Then they warned him to never speak of it again. Now, he will not speak about it with me (I am non Native). A bit frustrating, because I'm in it! But gotta respect his decision.

The reason they say not to talk about them aloud, is because it attracts both skinwalkers and wendigos. Also I believe it sounds like you are describing a wendigo, not a skinwalker, especially that far north.

There is a mental illness known as Wendigo that is mainly believed to be fake that's main trademark is a deep craving for human flesh.

I've always maintained these "new" descriptions of Wendigos and skinwalkers are inaccurate. Being native myself and knowing of, and maybe more importantly understanding the rationale behind such creatures, these overly dramatic descriptions just don't fit what I know of them. I've always wondered where these newer descriptions come from. I believe people are seeing these things, but the eyewitness accounts are vastly different from what I know of them. I worry that we may be inclined to try and protect ourselves from them using methods created for a completely different creature.

I absolutely believe in the wendigo.

Not really, because I haven't seen or heard of any groundbreaking evidence or had anything remotely close to an encounter. But at the same time, I would rather not go out and prove it.

While there doesn't seem to be any proof that the Wendigo doesn't exist, there are plenty of stories passed down from generation to generation claiming that it does. So it is hard to say for sure. I personally believe that it is indeed real, there is a lot in this world that we do not understand so it would not be completely impossible for the Wendigo to exist. That is my opinion. Plus every legend is based on some truth.

Well the cannibalistic disease is pretty real.

Well as you can see the Wendigo has mixed reviews by people if it really exists. Me, I kinda think they do exist, do you?

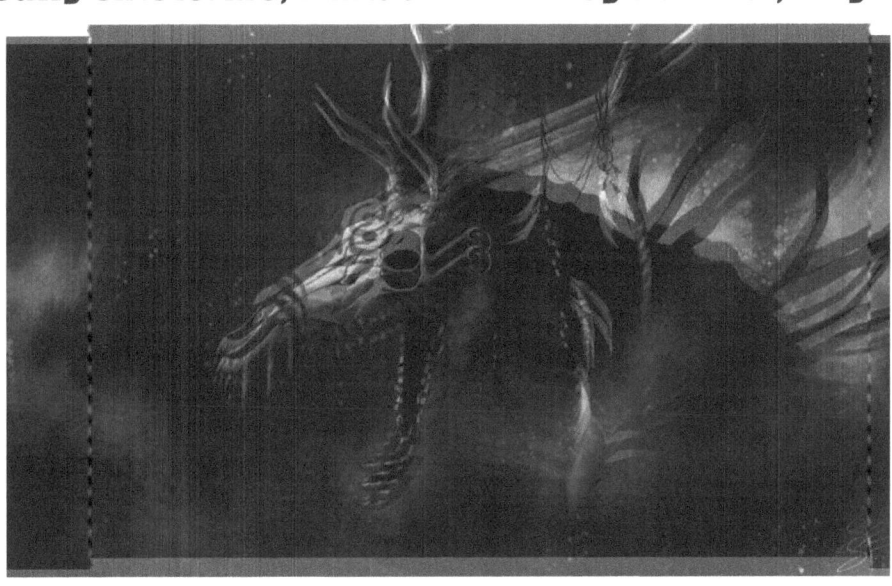

The Abominable Snowman or Yeti

The Yeti are a cousin of Bigfoot living in the Himalayas mountains. The creature is not really white in color. Reports of it being brown or reddish in color. And walks on 2 legs weighs 200 to 400 pounds and is about 6ft tall in height. The creature is known for being an omnivore which means it eats both plants and meat.

A Buddhist monastery in Pangboche Nepal has a hand of the Yeti. but the hand has never been tested by the scientific community. So I am still on the fence about it, really the hand of a Yeti.

So i reached out to social media to see if anyone believed in the Yeti i was shocked in the responses i received.

I personally like to believe they are real as it isn't too outlandish that a creature that is well equipped for frigid temperatures could live in such cold conditions.

But my other reason for believing is how wonderful a thought it is that there is a fluffy humanoid beast who lives in the snowy mountains. Just living his best introverted life.

Well to make the yeti an endangered species I have to bring in a specimen dead or alive so I don't want to kill the yeti. The Wendigo on the other hand is a killing machine and a danger to society, it's just a precaution.

Real. They live in too harsh of conditions.

Real. Both Yeti and Bigfoot exist but are not primates or any elusive species native to Earth. There is more to this than meets the eye, see Steve from how to hunt on YouTube for more information. Dig deep and you will find the uncomfortable truths.

Yes, unequivocally.

Absolutely! IMO they are descended from Himalayan bears! And evolving into the Yeti bc of sheer survival needs and isolation!

As you can see, the Yeti is a good debate on social media. I have seen Bigfoot to say that the Yeti is real.

Ghost ships of the Great Lakes

19

The ghost ships of the Great Lakes date back to the 1600's. When the old wood ships would sail across the lake to do trading. The Great Lakes can be dangerous as well in the past and today with storms and ice.

The story of Edmund Fitzgerald is a legendary tale and Ghost Story. There have been over 6,000 ships that have sunk in the Great Lakes.

Ships like SS Carl D Bradley that sank in Michigan in Nov of 1958. 33 crew members died in the sinking; only 2 crew members lived to tell the tale.

Le Griffon sank in Lake Michigan in 1679 it was the first ship to traverse the great lakes. All the crew had died. The Tv show Expedition Unknown did a story on it and with new cutting edge we might just know what happened to the lost ships and souls.

Well i went to social media for the Question are in fact these Ghost ships real and what people thought. I was very surprised by what I read and the overwhelming response I received.

Here are some of the best responses I have ever received.

Absolutely. I grew up on the Great Lakes and still live in the region. It's an area with tons of high strangeness. Ghost ships are just the tip of the iceberg.
There are sightings of bigfoot, dogmen, and UFO's. There are a lot of Native American burial mounds, which allegedly contain giants. Not to mention plenty of Missing411 cases. All in all, there are a lot of interesting stories in and around the Great Lakes.

The bannockburn is known as the flying Dutchman of lake Superior

There's definitely been a lot of tragedy on the Great Lakes, like the wreck of the Edmund Fitzgerald.

When I was in 1st grade, I was obsessed with everything surrounding Edmund Fitzgerald.

I saw something pretty unexplainable on the Hudson River once. A Lot of ships have gone down on the Great Lakes. So, I believe something like that is possible.

I live in the Upper Peninsula of Michigan, between Lake Superior and Lake Michigan. I've heard all kinds of things growing up between the two. Lots of ghost stories, ghost ships, cryptids and such.

I'm a Wisconsin native (born and raised), so I'm close to two Great Lakes, and there are a metric fuck ton of ghost stories about ghost ships. Not to mention Sasquatches, dogmen, UFOs, and other unexplained stuff. Wisconsin has its fair share of the paranormal, supernatural, and bizarre. We are the home state of the Beast of Bray Road, Hodag, and more. I wholeheartedly believe that the Great Lakes (and the area surrounding them) are hotspots for the unexplained.

Hell yeah. I've lived in the upper peninsula of Michigan right on the coast of Lake Superior my whole life. The upper peninsula is a fucking hotbed for strange activity

Whether you believe in Ghost or Ghost ships the Great Lakes are a paranormal hotbed of activity through the centuries. The waters of the Great lakes are a great conductor for paranormal activity.

The creature known as the Jersey Devil

Paranormal question of the day

Do you believe in the Jersey devil real or fake?

The Jersey Devil is a creature I think is miss classified. Some believe that the Jersey Devil is a Bigfoot in the Jersey Pine Barrens.

The first sighting of the Jersey Devil was in the early 1900's. But wasn't reported until 1909 when the creature sightings started coming in. The reports of a winged biped with hooves, but reports from local residents describe the creature in many different forms.

One of the most famous reports was from as early as the first settlers to the Pine Barrens. Whether you believe in the Jersey Devil that's for you to decide.

Went to social media to see what my peers thought about the Jersey Devil. I got some mixed reviews.

My daughter swears she saw it when she was about 3.

The creature, possibly. But not the one from the story of being born and flying out the chimney.

I agree, that story is a bunch of nonsense. I think it was made to demonize the Leeds family for whatever reason and somehow that story has ended up merging with the creature phenomenon that is probably unrelated. I say unrelated because nearby in Maryland there were reports of a similar creature the locals call the Snallygaster which derives from Schneller Geist or Swift Spirit but these incidents occurred in the 1600s much earlier than the Leeds Devil story started spreading. I think the two creatures are related or maybe even the same creature that goes by different local names. A recent camera trap photo of the Snallygaster looks just like the Jersey Devil.

Somewhat related, about 2 years ago there was a bat humanoid creature spotted all over downtown Chicago. Flying over the lake, seen standing in the park wrapped up with its wings, etc. Lots of independent witnesses over a couple weeks.

Holy shit I literally just started a lore podcast about the Jersey Devil 10 minutes. Decide to get on Reddit and look up cryptid stuff bc that's what I'm into right now and the first thing I see is the Jersey devil.

Well, it really depends. If you want something compelling, look up phenomenal week in regards to the Jersey Devil. You'll find some interesting info...

I actually live near the pine barrens and read books about this cryptid for a large chunk of my childhood. Not sure if it's real truth be told, however apparently sightings still come in. Maybe it is!

So the story goes, in the Pine Barrens lived the Leeds family – "Mother Leeds," her husband, and 12 children. She bore all 12 children without suffering any miscarriages or stillbirths, which seemed impossible back in the eighteenth century. People called her a witch. Her husband rarely cared for his children, preferring the bottle to his family. When Mother Leeds learned of her thirteenth pregnancy, she cursed her unborn child. "Let this one be the Devil," she shouted at the heavens.

If the creature is real why hasn't there been any photographic proof of the creature?

Is this creature just a made up folklore to scare little kids. Does our mind imagine these beasts up and tell us they are real? When it comes to the Jersey Devil I really think it is a folklore and not a real beast roaming the Pine Barrens of New Jersey.

The Legend Known as Loch Ness Monster

If you ever watched In Search of as a kid you remember the episode on the Loch Ness Monster. This along with Bigfoot has always made me believe that a dinosaur could be living in Scotland and Lake Champlain in New York.
Yes in the United States called Champ a relative of the Loch Ness Monster.

Let's do a little History on Champ and Loch Ness.
In Scottish folklore Loch Ness or Nessie in (in Scottish Gaelic: Uilebheist Loch Nis). A creature who said it's home to be Loch Ness in the Scottish Highlands.

The Creatures are described as both having long necks when they stick their heads out of the water. The first sighting of Loch Ness was on May 2, 1933.

The first sighting of Champ was in 1609 by the first Europeans. But in 1970 an issue of Vermont Life ran an article on Champ. Champ is also known as Lake Champlain Monster. You can find all kinds of merchandise in Lake Champlain and local businesses with champs names.

So with this one i decided to reach out on social media and see if anyone believes in champ or Nessie. Do you believe?

Everyone says they have seen it in different parts of the world but it could be an underwater sea creature who had babies and lived for many years who knows.

My grandfather fished Lake Champlain often. He was not one to make up wild stories, but believes my great uncle and he saw a large serpent type creature not far from the Champlain Bridge during the 1980s. This was the only story of this kind he ever told me. I wouldn't doubt him. It wasn't as large as some described. About 15-20 ft. They saw a silhouette of a couple of humps and a snake-like neck (6-8" diam) and head (see image) moving across the water about fifty feet from them. Not sure what they were seeing they moved the boat closer. The creature went under and reappeared further away. They approached it again. It went under again and that was the last and only time they saw it.

I believe there's something there.

On the edge about both.

I just don't know. I need to see it to believe it.

Well, we know the plesiosaurus existed and it's one of my favorite ideas that some dinosaurs survived extinction. Unfortunately other than old stories and I saw it with my own eyes, "eyewitness testimony being the least reliable", it seems it doesn't exist. The loch is under surveillance 24 on a live stream for a long time and no one has seen a thing. Alas I hope we find it still.

I am confident that Nessie is actually a Greenland Shark. Or I should say Nessie sightings have been sightings of Greenland Sharks. They Live near the bottom of Loch Ness, and are around 20ft. Long, have smooth skin that is dark gray to black in color, and on the rare occasion they do surface they just bob up then back down like a submarine. This would explain the Serpent like body people claim to see. Lastly Greenland sharks have poor eyesight and have been known to accidently attack boats thinking they're food. As for Champ, I'm not so sure. Sonar signals from a large unknown animal have been confirmed from that lake and we know it has no whales or dolphins.
I hope this helps explain what is really in our lakes and oceans.

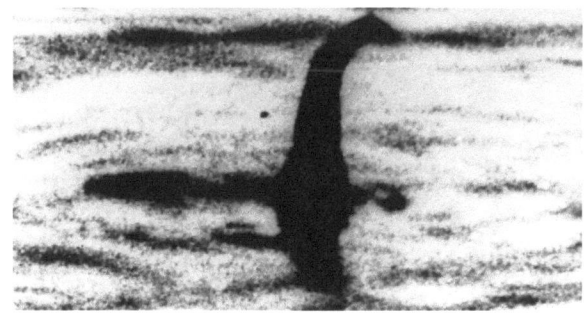

The MothMan

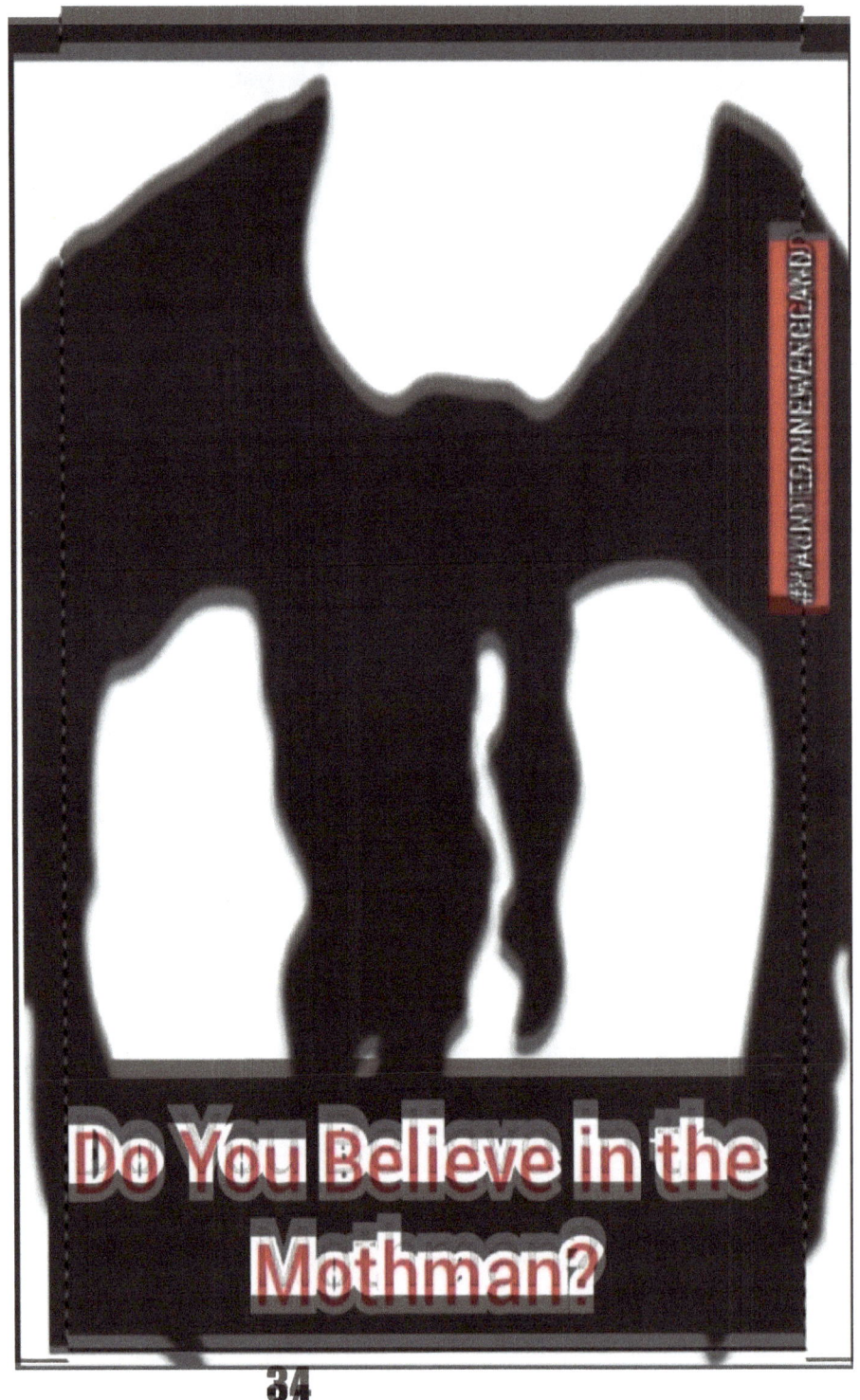

The Mothman the paranormal and cryptid. This one i have a hard time believing in for some strange reason this cryptid i kind of think is more a folklore and a paranormal story. I am not sure what I have seen in movies and online and books.

They all say that this creature has glowing red eyes and mouth like wings and has a body of a man. In West Virginia they hold festivals every year. Why is this creature so popular?

So in doing the history of the Mothman i found some possible truth about the creature.

......Seen in the Point Pleasant area on Nov 12,1966 to Dec 15, 1967.

,,,,,The first newspaper reported on the creature in 1966 in the Point Pleasant Register. The article was titled Couple see mansize bird.. Creature.

.....The national press picks up the story of the Mothman.

.....Reports start coming in shortly after the first report by many local citizens.

Ok after doing a history check I decided it was time to see what my peers had to say regarding the Mothman.

 I keep an open mind. I mean look back at the Egyptians, they had gods with bird heads, lion heads ect. I believe in aliens so anything is possible. That's why I research everything.

 went through a similar research over the thunderbirds--seems there are other species appearing on our planet that are not extinct as we once thought such as loch-ness,yeti, and many others i try to keep an open mind and investigate or research as much as possible but to answer if they are real--i don't know,could they be other things possibly--mysterious as all get out.

 believe anything can be true.

More than likely a Harpy Eagle.

"In this day and age, a claim about the Mothman, people would just scoff at it," "People are now skewed to not believe anything, even if it's real."

Do we know what's really out there? Is the mothman real? The Mothman could be as real as you and me , you decided.

THEY ARE CALLED SHADOW PEOPLE

PARANORMAL QUESTION OF THE DAY

DO YOU BELIEVE IN SHADOW PEOPLE?

REAL OR FAKE

This one i can say without a shadow of doubt there are shadow people.

I Have been investigating the paranormal for 30 years. And I have seen ghosts up and down the east coast. When you see a shadow person it looks like nothing you have ever seen is absent of light. Some believe that these are spirits that are stuck in limbo.

My first encounter was at the House of Seven Gables in Salem Ma. I was on a field trip with my 6th grade class and we had a tour at the house when I decided to stay behind and touch some objects. As i was touching a dish i saw a Shadow go from left to right, now remind you this is during the day. As I saw the shadow I felt a icey hand touch my shoulder and heard "Get Out" and I ran like a little school girl back to the bus.

As you can see i believe in shadow people so i decided to reach out to social media on this subject and here are some of the responses:

I "saw" these at Waverly Hills Sanatorium in Kentucky. I am a skeptic, so I feel as if they were tricks of the light due to the way the hallways and windows let light in. But, my friends were very disturbed by what they saw. One friend of mine had feelings of anxiety and negative thoughts for several weeks after our visit.

What is the likelihood that thousands of people who have never heard of sleep paralysis or shadow people all experience an event where they can't move or speak and see dark humanoid figures standing over them or around them?

Conversely, what is the likelihood that some doctor or psychiatrist starts telling people "your brain is stuck in a sleep state even though you are conscious and those shadow people you are seeing are just a figment of your imagination?
Keep in mind that a doctor is a person of science and would be looking for a scientific answer to the problem, and would ruin their careers if they were to say that it is actually something paranormal happening.

They may look at brain scans and say oh well this area is active so it must be your brain making this happen, but what if that is only the symptom and not the cause?

I've never had sleep paralysis but my best friend did for several years in our teens. I have however seen shadow things while wide awake.

As always, trust your gut. That sixth sense that tells you something isn't right, is usually correct.

I have never encountered one or known anyone to have, but at the same time I have not opposed their existence.

I've had multiple experiences from childhood to present day, starting as a shadow on the wall to a more recent live figure rising from the ground. I believe in them highly.

Seen them more than once, definitely real to me.

I've had multiple experiences from childhood to present day, starting as a shadow on the wall to a more recent live figure rising from the ground. I believe in them highly.

When I was a kid, around 5ish, I lived in this house with a wall of white closets. I remember seeing a shadow person/figure who looked like a kid around my age with super curly hair. I yelled at him to go away, and he disappeared between the cracks of the closet doors. I believe!

possibly- i've maybe seen them a couple times while it's dark in my room, but it could also be just my eyes playing tricks on me. As you can see that people have had paranormal experiences with Shadow People very real encounters. Do you believe now?

Native American Thunderbird

Paranormal question of the day
Is there such thing
As the
Thunderbird?

The Native American described the Thunderbird as a supernatural being, the enormous bird was a symbol of power and strength that protects humans from evil spirits.

It was called the Thunderbird because the flapping of its powerful wings sounded like thunder, and lighting would shoot out of his eyes.

The Thunderbird brought rain and storms, which could be good or bad. Good - when the rain was needed or bad when the rain came with destructive winds, floods,and fires caused by lighting.

The bird was said to be so large, that several legends tell it picking up a whale in its talons. They are said to have bright and colorful feathers , with sharp teeth and claws. They are said to live in the tallest mountains.

Whether the large bird is really as big as they say it is, why haven't we found remains of such animals? So I decided to go to social media and see if the Thunderbird was real or not.

I think in the native culture there is in one form or another.

I don't believe it. I think it originated as a mistaken identity and a fascinating legend was thus created. This could be said for many things paranormal or cryptid.

As you can see the public is kinda on the fence if the Thunder bird is real . I want to see hard data on this creature. But I do believe IN Native American folklore.

You be the judge

Do you Believe in Vampires

PARANORMAL QUESTION OF THE DAY

DO YOU BELIEVE IN VAMPIRES?

#HAUNTEDONNEWENGLAND

Vampire stuff of movies and folklore. I admit I am one who was sucked in as a kid the movie with Bella Lugosi I thought was real . Vampires are real in many cultures and many countries as well.

When i started my research for this topic i was amazed that Vampires were used as treatment for Tuberculosis during the civil war time.

So I set out on social media to find out people believe in Vampires or not, here what they say :

Yes energy vampires can drain you dry.

Honestly, if you've ever smelled blood that's been through the digestive system, either in vomit or feces, trust me, you'll hope no one actually tries to digest it. I'm a nurse with a pretty strong stomach and that smell is one I don't EVER want to smell again.

Depends on what context you mean. Paranormally, not likely. Historically, that's another version.

psychic vampires yes, which feed on various human energies.

The question is Do vampires believe in you?
For the record though, I do believe vampires exist in several forms.

The Hollywood sort no. Vampires yes, they are not immortal, don't bite people, don't turn into bats.

They are people like Father Sebastian, Sanguinarius Lady CG and many others.

No, I only think the people of today just dress the part teeth included, and they like the taste of blood. They all get excited for the awards they think they win from their academy award performance.

I once thought I did. Then I worked out the logic. Not possible.

I believe at one time they existed.

Not sure, highly unlikely though.

I wish they did.

As you can see that some people believe in different kinds of vampires. Some believe in the movie and the romance of vampires. Do you think that vampires are romantic and sexual beings? The legend of the vampire should be told in a way that explains the history and not in pop culture.

Do you believe in Demons

Paranormal question of the day

Have you ever encountered A demon?

Demons are a very strong subject when it comes to the paranormal. I have always said that investigating demonic or inhuman haunts are like winning the lottery. Until I had a case in Vermont that changed my mind . When i witnessed a demonologist scratched across the face and bloody. After that case the demonologist left the paranormal field.

The word demon means an evil spirit or devil, especially one thought to possess a person or act as a tormentor in hell. Demonic entities in the Old Testament of the bible. Well I wanted to see what my paranormal friends thought about demons. I was surprised by my findings.

We all have but many just don't realize it . They are in our lives every day. It's called good and bad . Remember the main demon alias the devil is a deceiver

I believe so. The most interesting part of it was several years later, when I told my mom about it, she said the same entity attacked her when she was about the same age. My mother is a physician and is a very scientific and practical woman, so I was amazed to hear this from her and even more shocked that she didn't poo poo me.

yep...had to turn the case over to a demonologist.

I encountered a demon in fall 2015 when I, being none the wiser, went to a bridge in PA at night and was about to get out of the truck when I saw red eyes staring at me. I sat still in my seat, frozen as a wave of nausea swept over me. I told my grandfather (who was with me) what I was seeing and had to leave. He obliged & drove as quickly as he could to get away from the historic yet haunted site, only for me to realize too late that the demon followed us home. While evil left my grandfather alone, it tormented me until I was easy enough to possess not once but twice in a week. I remember my possession but vaguely. If it wasn't for a certain angel coming to the rescue, I shudder to think what could have happened to me.

I barely escaped the demon with my life that year. Demons, like ghosts, are drawn to people that are gifted with psychic abilities. While I love my gifts as a medium and sensitive, I don't like being connected to evil. Responsibility and a lot of it comes with this gift. Most of the spirits I come across are nice but others are not, and rarely do I run into the demonic. I long to write a book about my numerous paranormal experiences.

Yes I am not sure from where but native American was tied to it.A door was open to someplace and it controlled all areas spirits trapped there. Was swallowed into its thick non light penetrating darkness. Was controlled and always mad. Guess once cleaned of it came to an understanding of a kind as it cost us health and home. All ok now and still sometimes re-investigates sight but somehow I know it will not attack me again. Maybe respect maybe already did what it wanted. Not sure but leaves me alone now.

Yes i help people remove them i never ever communicate with them no point!

Yes I certainly have...I banish them from others.

As you can see my fellow investigators have strong convictions when it comes to demons. Have you encountered a demon or a spirit ? Let me know how much I love to hear from you.

The Legend of the Bubak

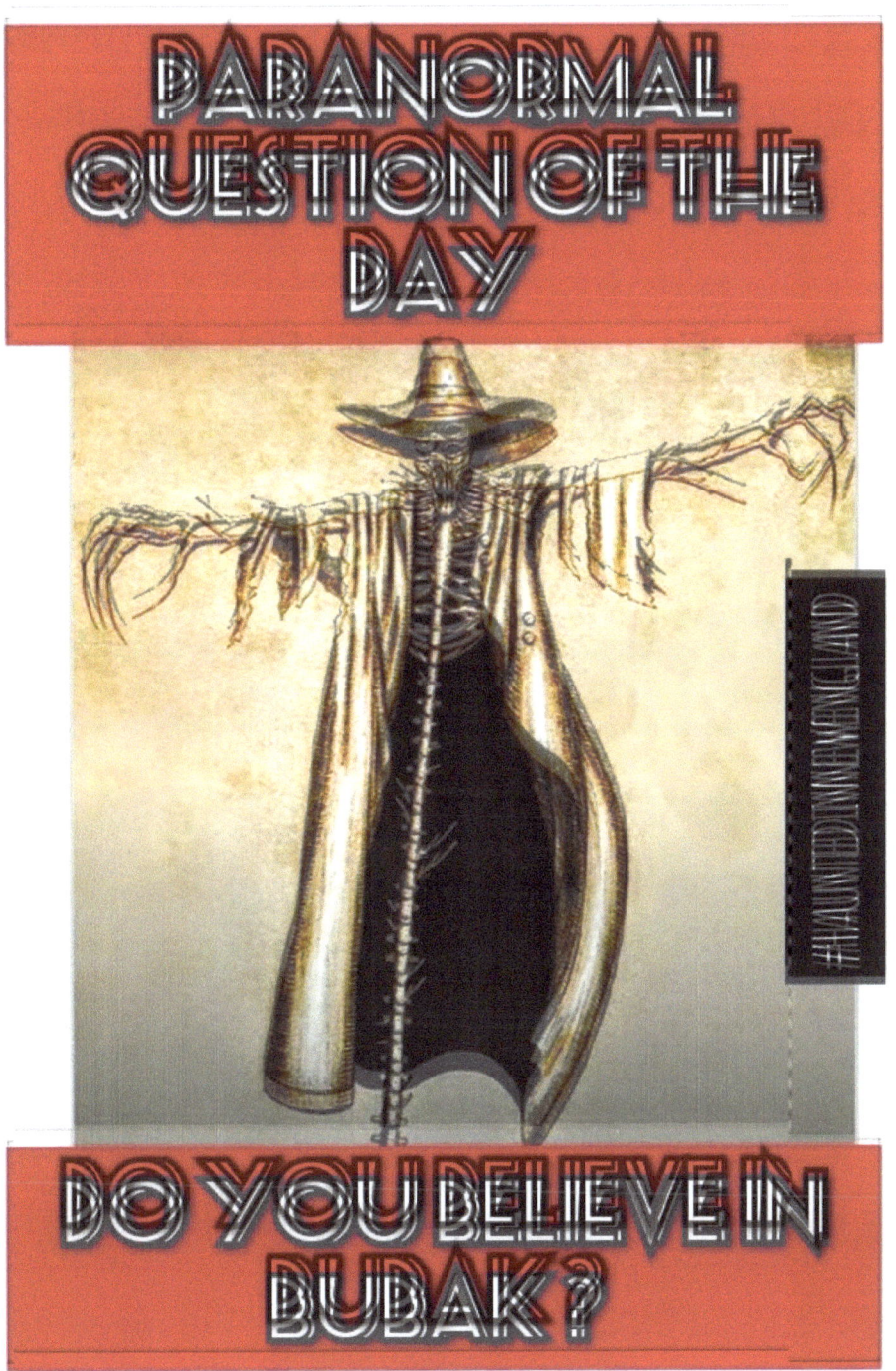

When it came to the Bubak I was skeptical about this creature being real. But after doing research I found out it was true. The Bubak is also known as the sack man, or in my eyes scarecrow. The story of the Bubak dates to the 1600's to the 1800's. The **Sack Man** (also called the **Bag Man** or **Man with the Bag/Sack**) is a figure similar to the bogeyman, portrayed as a man with a sack on his back who carries naughty children away. Variants of this figure appear all over the world, particularly in Latin countries.

n Armenia and Georgia, children are threatened by the "Bag Man" who carries a bag and kidnaps those who do not behave. In Hungary, the local bogeyman, the *mumus*, is known as *zsákos ember*, literally "the person with a sack". In Poland children are frightened by the *bebok, babok, or bobok* or who is also portrayed as a man with a sack. In the Czech Republic and Slovakia, a similar creature is known: *bubák*. It's a creature without a typical form, connected with darkness or scary places, making children fear but not taking them away usually.

The character of *čert*, the devil, is used for that instead ("Don't be naughty or čert will take you away!"). In Russia, Ukraine and Belarus, *buka* ("бука"), *babay* ("бабай") or *babayka* ("бабайка") is used to keep children in bed or stop them from misbehaving. 'Babay' means "old man" in Tatar. Children are told that "babay" is an old man with a bag or a monster, usually hiding under the bed, and that it will take them away if they misbehave (though he is sometimes depicted as having no set appearance). In Turkey, Kharqyt (Turkish: Harkıt means "Sack Man"- also called *Öcü, Böcü* or *Torbalı*) is portrayed as a man with a sack on his back who carries naughty children away to eat or sell them.

So I decided to reach out to social media to get their thoughts on the Bubak. This is what I got for responses.

I believe in the belief of the individuals in the infected area. These things can take on the whatever form they chose it seems

I've never heard of this one

Whether you believe in it ; it is your call:

Haunted Forests

#HAUNTED

Paranormal question of the day

DO YOU BELIEVE IN HAUNTED FORESTS?

Why do we have fear of the woods, is it the fear of the unknown ? As a kid I remember playing in the woods of New Hampshire. I never thought the woods were haunted, Boy was I wrong.

After doing research on Haunted Forests I found out throughout the world there are many of them in the world.

Aokigahara Forest

Sitting on the northwestern side of Mount Fuji, Aokigahara Forest (more commonly known as "Suicide Forest") is the definition of tragic beauty. Sometimes referred to as the Sea of Trees, it has been the site of numerous suicides, dating all the way back to the mid 1900s.

Isla de las Munecas, Mexico

As if forests weren't scary enough on their own, this one located along the canals of Xochimico, near Mexico City, is covered in dolls that hang from the trees. Local legend has it that the island's caretaker hung the first doll to honor a little girl that he had found drowned on the island (the doll was thought to be hers).

The caretaker then felt as if he was being haunted by the girl's spirit, so he continued to string up dolls in hopes of appeasing her. Fifty years and plenty of dolls later, he reportedly drowned in the same spot the girl did.

Today, the island has become a popular tourist destination. Visitors claim that the dolls move their eyes, heads, and limbs, and that they're possessed by spirits.

These are just a few locations that are known to be Haunted or have strange creatures lurking deep in the woods. So I reached out to social media to see what everyone's thoughts on this topic here are the responses.

Absolutely I've had the pleasure of investigating the Hoia Baciu Forest in Romania. It did not disappoint.

Freetown forest is a testament to that.

Parents took me through the Ardennes..tracing our uncle Bea's route in ww2. I wasn't comfortable. I was really young at the time, 10 years old. At night..i could see, hear and smell the events. Lost that ability as I grew older. Thankfully.

Hockomock swamp.... another one.

MY FINAL THOUGHTS

As I decided to do the research on this topic it was going to be pretty clear what i was going to write about.

In 2014 i wrote my first Paranormal Questions of the day book 1, i never thought i would write a book 2 7 years later.
In 2014 I won the IPAA award for education in the paranormal field and considered the paranormal originator of the paranormal question of the day.
The cryptid field I started to fall in love with and started to do research and it peaked my interest. Do I believe in Bigfoot? I do , the Mothman not so much. It has been my pleasure to meet people who talk to me about the paranormal and Cryptids as well. I hope this book helps you in some way to believe in the unknown.
REMEMBER WE LIVE IN A GREAT BIG WORLD GO AND EXPLORE.

AUTHOR ERIC PERRY

Find these other great titles by Eric Perry

BIGFOOT MYTH OR LEGEND

FARMS IN NEW YORK , PICTURE BOOK

HAUNTING OF HIGH ROCK TOWER , CASEFILES OF

THE HAUNTING OF RANSOM O. GORE

REAL NEW ENGLAND HAUNTS TWO

PARANORMAL QUESTIONS OF THE DAY BOOK 1

TRUE HAUNTINGS IN NEW ENGLAND

You can find these great books at Barnes & Noble, Walmart.com, BAM!, and Amazon and Kindle

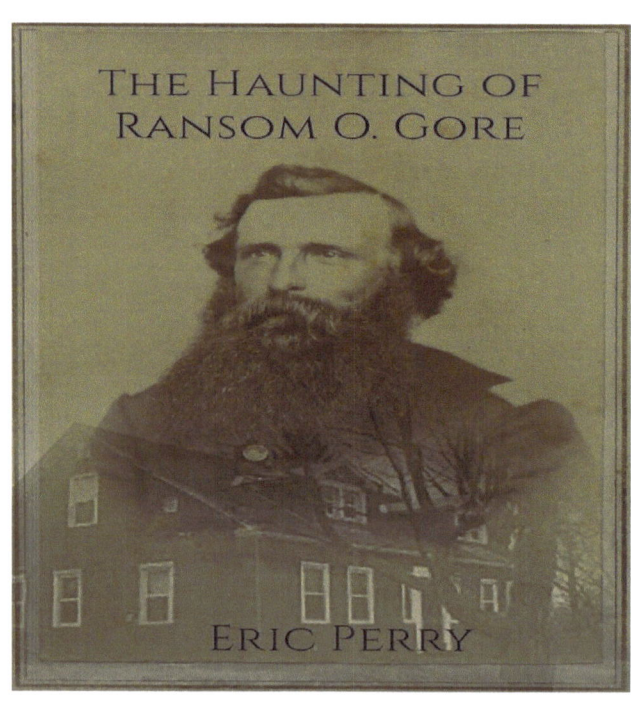